Hexagonal Water
The Ultimate Solution

Discover why drinking the liquid crystal known as hexagonal water can provide superior hydration, enhanced nutrient absorption, more effective detoxification, increased metabolic efficiency, and improved cellular communication.

Revised edition 2007

MJ Pangman, M.S.

UPLIFTING PRESS INC.

Revised edition March 2007
Cover Design: Jenique Graphic Design

ISBN# 0-9752726-1-6
Printed in the United States of America

The information in this book is for educational purposes only. It is not intended as a substitute for medical advice, nor is it intended to replace the counsel of a health professional. Neither the publisher nor the author directly or indirectly dispenses medical advice. If you have a health concern, contact an appropriate health professional.

Contents

Preface

As a writer in the field of complementary medicine, I have been introduced to many of the latest discoveries in health. But water? Like most of us, I knew it was important to drink pure water, but I had no idea that water's *structure* could be modified. Consequently, the concept of "hexagonal" water was a complete surprise.

Initially, I tested a structured water product using methods that measure the body's response to various forms of energy (biofeedback). Those tests were entirely positive. In fact, they were so positive that the only way to describe the measurements we took was "off the charts." When I began to drink the water, I experienced a noticeable improvement in energy and several indicators of cellular detoxification.

Feeling like I was onto something, I began to research the concept. I discovered that there was abundant evidence for water having structure. I also discovered that the structure of the water immediately surrounding DNA and other macromolecules was highly ordered (structured). But could a whole glass of water hold a specific structure for an extended period of time? It appeared nearly impossible to prove with our current level of technology. At this writing, we do not have methods which allow us to actually "see" molecular structure.

During my search for more information, I found references to a Korean book written by the renowned scientist Dr. Mu Shik Jhon. During his 40 years of extensive research, Dr. Jhon helped establish the concept of structured water, or as he came to call it, "hexagonal water." Unfortunately, there was no English translation of his book. Then, in 2003, through the manufacturer of a structured water-making device, I was offered the opportunity to work with Dr. Jhon to edit and translate his book (now published in English as *The Water Puzzle and the Hexagonal Key*). The process was an eye-opening experience. His book gave authenticity to the term *hexagonal water.* It also provided the results of limited testing which showed that consuming the hexagonal water made using methods he identified, had positive effects on the human body.

Dr. Jhon passed away shortly after the English translation of his book was published, but his research and vision have given birth to further study and to a number of "hexagonal" water products. We have recently learned to "imprint" water with subtle energetic frequencies, and we are not far from a day when water will be used as a primary medical intervention.

However, many scientists remain skeptical because of the lack of enough "hard science." That too, will come. In the meantime, there are many healthcare professionals who have used a variety of energy and biofeedback testing methods to validate the efficacy of consuming hexagonal water. These methods show that hexagonal water is highly beneficial for the human body – capable of making a profound difference in the health and well-being of those who drink it. Furthermore, the growing number of health professionals who *use* some form of hexagonal water as an adjunct to their therapies is a remarkable testament to its efficacy.

Of course, all hexagonal water products are not equal. Some are better than others; all do not produce the same results. As with anything, the consumer should use discretion in selecting a hexagonal water product or hexagonal water-making device that is of the highest quality.

MJ Pangman

Hexagonal Water

The Ultimate Solution

INTRODUCTION

Today it is common knowledge that dehydration is at the root of many diseases, that aging is a function of cellular water loss, and that 50-75% of Americans are chronically dehydrated. Thousands of books have been written about water in an attempt to establish the "facts," but the simple fact remains: we have just begun to get our feet wet at the fringes of an ocean of information about the amazing substance that governs life on this planet.

Pioneering research is confirming that water is not simply water – no matter how pure. Our scientific development is finally allowing us to explore the hidden qualities and characteristics that make water entirely unique – the qualities that give water "life" and "energy." One of these qualities is its "structure."

Hungarian biochemist and Nobel Prize recipient (1937) Albert von Szent-Györgyi stated, "Water is life's mater and matrix," (mother and medium). He recognized that water was the beginning and the sustenance of life, and he recognized that there was much more to water than the scientific community was aware of at that time. Dr. Szent-Györgyi was an active proponent of the study of water. He participated in and encouraged much of the early research regarding the structure of water.

Dr. Szent-Györgyi also stated: "Discovery consists of seeing what everybody has seen and thinking what no one has thought." His statement is perfectly poised at the beginning of a discussion on hexagonal water. It describes the current scientific view of water as "seeing what everyone has seen" and the possibility for understanding a bigger picture.

By its very nature, the scientific community is often the last to acknowledge new ideas and developments. Even though mainstream science has not yet embraced the concept of hexagonal water, a flood of research is beginning to support its existence. The following pages provide evidence for the existence of hexagonal water as well as for its potential for superior hydration, enhanced nutrient absorption, more effective detoxification, increased metabolic efficiency and improved cellular communication.

Water Basics

Water is much more than a solvent or delivery system to carry nutrients throughout our bodies. Water is the essence of life itself. It is the highway that delivers electromagnetic currents to fuel our cellular reactions. It is the resonance field through which our cells communicate. Water is involved in the transfer of data from DNA and in the transmission of vibrational information.

Water transports, stabilizes, signals and hydrates. But, what we have just begun to realize is that the specific structural form of water referred to by Dr. Jhon as "hexagonal" water contributes to more *efficient* transport, *greater* stabilization, *clearer* signaling, and *superior* hydration.

Water quality – more than purity

Millions of dollars have been spent to remove the physical and chemical pollutants from our drinking water. However, even the purest water may still contain *vibrational* pollution. It has been said that water is the *most programmable substance* on the planet. This is because water has the unique ability, referred to as water's "memory," to retain the energetic signature of whatever is placed in it. Water even appears to have the unique ability to store and transfer this information within the human body.(3)

Water that is energetically polluted can be equally as harmful as water that is chemically polluted. The more times our water is recycled, the more energetically polluted it becomes. Since water is involved in every biological activity, and since we are between 50 and 90% water, depending on our age and health, it is obvious that when the water we drink is not clean (chemically and vibrationally), it can have a major impact on our body's ability to function properly.

One other factor is emerging as a major element in the quality of the water we drink – molecular structure. Molecular structure, the organization of water molecules in a repeating geometric pattern, appears to be equally as important as the chemical and vibrational condition of the water. Molecular structure appears to play a remarkable role in the movement of water throughout the body (hydration), as well as in the crisp, clear transfer of energy and information.

The anomalies of water – explained by water's structure

Water is a unique substance; it does not fit the pattern of other liquids. In fact, there are over 40 distinct anomalies of water[6] – differences that give it the ability to sustain life. Scientists have discovered that these anomalies can be explained in terms of water's structure. Physical properties such as heat capacity, compressibility, viscosity, dielectric constant, and surface tension all change with temperature in ways that indicate structural changes in the water.

The most obvious of water's anomalies are noted as water changes temperature – a reflection of a changing structure as water approaches freezing temperatures (crystallizes). Like all matter, water contracts as it cools. However, just before freezing (4°C), water begins to expand. At atmospheric pressure, water expands 9% as it freezes. This is why pipes often burst in the winter. It is also the reason that ice floats rather than sinking to the bottom of a glass of water. This unique property insures that the freezing of lakes and rivers begins from the top down, providing insulation and protection for the life below. This trait is also the most obvious indicator that water has structure – a crystalline structure that develops as water freezes.

Water has a large heat capacity – far beyond other liquids. In other words, it has a tremendous capacity to store and transfer energy. The unusually large heat capacity allows water to absorb a remarkable amount of heat with a minimal change in temperature. This is why temperatures along coastlines are more moderate. It also explains why, being approximately 70% water, we are able to withstand large temperature fluctuations without freezing or overheating. Water also releases a considerable amount of heat as it freezes. This is why many farmers mist their crops to protect them from damage during frost.

Dr. Mu Shik Jhon explained why structured water could have

a higher specific heat value than normal water.[17] His work demonstrated that hexagonal water would have an even greater potential to perform work, to absorb temperature changes, to expel wastes, and to protect the body from energetic disturbances.

The hydrogen bond

When an oxygen atom combines with two hydrogen atoms to form water, the electrons from the hydrogen atoms spend more time around the larger oxygen nucleus than around their own hydrogen nuclei. This results in what is known as a polar molecule – one that has a positively-charged side and a negatively-charged side. Chemists usually depict the di-polar water molecule with a "V" shape, determined by the predominant angle of the hydrogen atoms in relation to the adjoining oxygen atom. The accepted angle of the "V" in a free water molecule is 104.5 degrees. (Fig. 1)

Because water molecules have a positively-charged side, they are attracted to the negatively-charged side of an adjoining water molecule. This forms what is called a "hydrogen bond." Hydrogen bonds are responsible for the formation of a network of interconnected water molecules that provide structural qualities to water. Hydrogen bonding accounts for all of water's so-called anomalies.

It is generally accepted that a greater degree of hydrogen bonding indicates a greater number of interconnected units, and a greater degree of structure to the water. At any given moment in time, random groupings of interconnected water molecules may consist of as many as 20 to 90 individual H_2O units. The extent of the hydrogen bonding depends on circumstances like temperature and pressure. Unique energetic circumstances help create the structural form of water referred to by Dr. Jhon as *hexagonal* water.

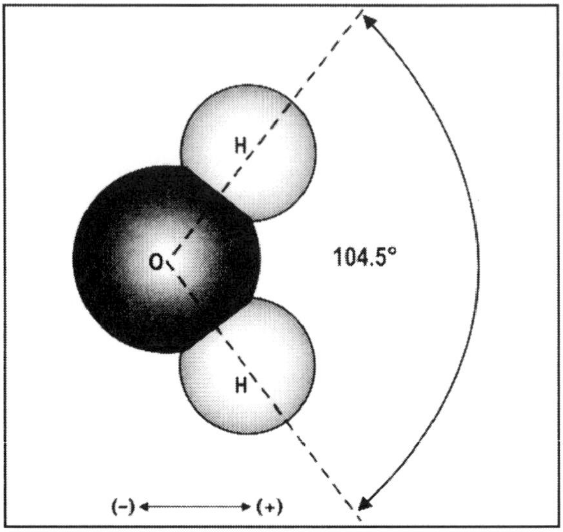

Fig. 1 The water dipole.
Water has a positive and negative side based on the predominant position of the hydrogen atoms in relation to the oxygen atom.

WHAT IS HEXAGONAL WATER?

Under normal circumstances, hydrogen bonds between water molecules continually break and reform. In fact, water molecules can exchange partners as often as every picosecond (trillionth of a second). Therefore, it is important to understand that the term "structure" has a unique meaning with reference to water. Structure implies a fixed position which is never the case in the molecular world. Regarding water, the term structure refers to a tightly hydrogen-bonded network that is more organized than normal water. The term hexagonal water refers to the specific crystalline organization generally accepted to exist only in the solid state (ice).

Although water molecules are in continual motion, some circumstances create coherent patterns of hydrogen bonding that exist for extensive periods of time.[(15)] Longer structural stability has been measured in the presence of certain minerals and proteins where the water immediately surrounding these structures has been said to be "as though it were in a frozen state."[(16)]

Hydrogen bonding creates the possibility for an infinite number of molecular arrangements in water. Numerous geometric structures have been identified. The most stable of these structures are the pentamer (composed of 5 hydrogen-bonded water molecules) and the hexamer (composed of 6 water molecules). When groupings of hexamers join together, an ice-like, liquid, crystal is established. Research conducted by Dr. Jhon showed that the formation of this hexagonal network within water had tremendous biological significance.

The difference between normal water (a structural conglomerate) and hexagonally-structured water (an organized crystalline matrix) is similar to the difference between a piece of quartz and a quartz crystal. Although they are chemically the same, a piece of quartz is more structurally random with an almost opaque appearance. Quartz crystals are organized in perfect geometric symmetry; they are "crystal" clear.

Quartz crystals are useful in the computer and timekeeping industries because of special piezoelectric properties which non-crystalline quartz does not possess. Computer chips and quartz watches rely on these piezoelectric (pressure-sensitive electric) properties produced by the geometry of crystalline structure, for accurate information transfer.

In water, structural stability is the result of *straighter, stronger* hydrogen bonds.[(4)] Stability also results from the mutual support and cooperation rendered by adjoining water molecules as the water forms a cohesive, coherent pattern. In essence, when hydrogen bonding increases, the cohesiveness of the resulting water also increases, creating a greater degree of structure and a more stable, more "open" water network.

The open water network established in hexagonal water results when the angle of the "V" (between the oxygen and hydrogen atoms) widens to create the 3-dimensional pattern. Rather than a 104.5 degree angle, the molecular angle in hexagonal water measures 109.5 degrees. This is considered to be the "perfect" tetrahedral angle for water where each water molecule serves as the donor and the acceptor of 2 electrons. This angle and the resulting geometry give rise to a hexagonal network – the same molecular arrangement found in ice.

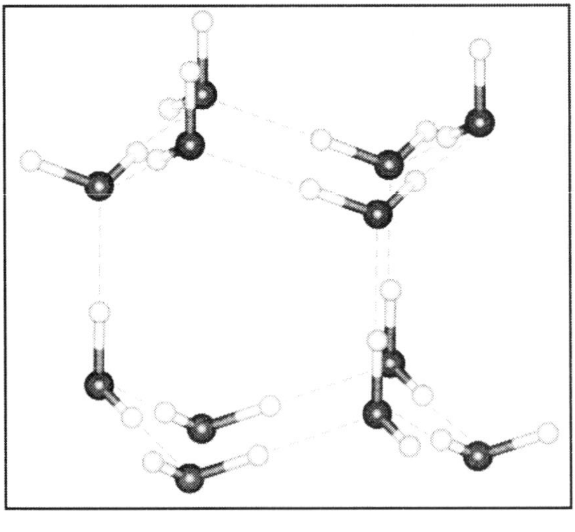

Fig. 2 The water hexamer

Normal water at 50°F is about 20% hexagonally structured. However, the influence of a weak magnetic field can increase hexagonal structuring up to 80%.[17] Certain other energetic influences are also known to produce greater structuring in water. These include: electric fields, lower temperatures, the presence of certain minerals, and some vibrational frequencies. The presence of any or all of the above factors encourages groupings of six water molecules which form the cohesive geometry referred to as hexagonal water.

Hexagonal water is *liquid crystalline water*. It is theorized to have coherent characteristics where electrons spin together and information transfer is almost instantaneous. Hexagonal water possesses a higher energy state – the key to providing rapid hydration, more efficient energy transfer, and the enhancement of numerous metabolic processes as outlined by Dr. Jhon.[17]

Energy potential

A majority of the water on the surface of the planet is a mixture of hexagonal units separated by other structural units and a tiny number of free water molecules. Even though water has a high energy potential compared to other liquids (the result of hydrogen bonding), it is no match for the energy potential of crystalline (hexagonal) water.

When water molecules join to form the hexagonal matrix, hydrogen bonds become stronger. And since energy is stored in the hydrogen bond itself, a greater degree of hydrogen bonding increases the potential energy. The energy is additive – just like increasing the voltage by adding batteries in series – the greater the hydrogen bonding in the water network, the greater the potential energy in the solution.

Verifying hexagonal water

The Lawrence Berkeley National Laboratory is one of few laboratories that have been able to physically view (via scanning tunneling microscopy) the existence of water hexamers.[24] Other methods "point" to structural differences in water samples.

Physical properties

Cold water has a greater degree of hydrogen bonding – an indication of its increasing structure as it approaches crystallization at freezing temperatures. Many of the physical properties of water show changes with temperature. Often these changes can be linked with structural changes (e.g. viscosity, density, compressibility, thermal conductivity, specific heat, dielectric constant, surface tension). If these physical properties change without varying the temperature or pressure, there is evidence that the structure of the water has changed. Measurement of these properties can provide helpful indications of water's structure.

Viscosity

Viscosity is the measure of a liquid's internal resistance to flow – sometimes thought of as thickness or stickiness. Viscosity is dependant on the forces which hold molecules together. Tightly hydrogen-bonded (cold) water molecules exhibit greater cohesiveness as evidenced by an increase in viscosity. At room temperature (20°C) the viscosity of water is 1.0 cP. At 5°C water's viscosity is 1.5 cP.

Density

One of the most well-known anomalies of water is the density anomaly which accounts for the fact that ice floats. As most liquids are cooled, the molecules approach each other more closely; density increases. Water is different. Its density increases as expected until just before it freezes – at which point (4°C) it begins to expand. At this temperature, the degree of hydrogen bonding (and subsequent structuring) overcomes the effects of cooling such that the density decreases; water expands as it forms a crystalline network. At 4°C the density of water is 1.00 g/cm-3. On either side of this temperature the density decreases. Hexagonal water, which has also been called "low density" water, has a density of .96 g/cm-3 (measured around macromolecules).[(43)] This tiny difference is a reflection of the more open structure caused by straighter, more organized hydrogen bonding. Hexagonally-organized water molecules are characterized by four neighboring (hydrogen bonded) molecules. Normal water at ambient temperature has five close neighboring water molecules[(23)]; and these are not necessarily hydrogen bonded.

Compressibility

Compressibility is a measure of density in response to pressure. Since increasing pressure breaks hydrogen bonds and collapses the structure of water, structured water is able to be compressed to a greater degree than normal water. This is illustrated by the variation in compressibility of water at different temperatures (45.9 units at 20°C vs. 50.5 units at 2°C).

Thermal conductivity

Thermal conductivity, the ability to conduct (transfer) heat, normally decreases in water as temperatures are lowered. Interestingly, thermal conductivity in ice is over 3 times that of normal water. The reason for this lies in the crystalline structuring of the molecules. In a recent unpublished study, computer simulation showed that by placing water in an electric field (which causes the molecules to align forming the ice-like crystalline matrix) the thermal conductivity potential was raised 3-fold.[(11)] This demonstrates the significance of molecular order as it relates to energy transfer.

Specific heat

Specific heat refers to the amount of heat necessary to raise the temperature of a substance by 1 degree. Water has one of the highest specific heat values known. Once again, this is due to the degree of hydrogen bonding. When water is heated, the energy is used first to bend and break hydrogen bonds. This does not immediately raise the temperature – it merely reduces the structure. Additional heat is then required to actually raise the temperature of water. Cold water (more structured) has a higher specific heat value than warmer water; it has the ability to hold more energy and to resist energetic changes to a greater degree because of its extensively hydrogen-bonded network.

Dielectric constant

Dielectric constant is a measure of the degree to which a substance resists alignment in an electric field. Because water is a polar molecule, one might think it would have a low dielectric constant (low resistance to alignment). But, because of hydrogen bonding, water actually has a high dielectric constant. In fact, the greater the hydrogen bonding, the higher the dielectric constant. This has recently been correlated with the degree to which water is structured.(26)(38)

Surface tension

Surface tension is defined as the measurement of attractive (cohesive) forces which are present at the surface of a liquid. Since water has a high degree of hydrogen bonding and its molecules are held tightly together, water has a high surface tension. This is the reason water "beads up" on smooth surfaces. Any factor that increases the strength and number of hydrogen bonds will also increase the surface tension. For example, cold water (with a greater degree of structure and correspondingly stronger hydrogen bonds) has a higher surface tension than room temperature water: 75 dynes/cm at 5°C vs. 72.7 dynes/cm at 20°C.

This may appear to contradict some theories which state that reduced surface tension provides for better hydration. However, when one considers that higher surface tension produces a stronger capillary effect (the ability to move against gravity into tiny areas) the improvement in hydration with hexagonal water is easier to understand.

Spectroscopy

Every molecule or substance has its own oscillation pattern or vibration which can be measured in electromagnetic wavelengths by various forms of spectroscopy. Water sends out different energy signals depending on the movement of individual molecules. When these signals are recorded, the picture is like a map.

In an interview, the renowned German physicist Wolfgang Ludwig asserted that in the future we would determine water quality not only chemically, but even more importantly, by spectroscopy.[21] According to Ludwig, the frequency spectrum is a physical copy of the geometrical structure of water. His experiments were able to show spectral differences in water samples[22] and he was able to transfer frequencies in water.[21]

NMR

Numerous researchers agree that NMR is a valuable tool for examining water's structure.[17][36][39] Since water is a polar substance (with a positive and negative side), it can be aligned in a magnetic field. Nuclear magnetic resonance (NMR) measures the time it takes molecular objects to return to their original position after magnetic alignment. The amount of time this takes is related to the size, shape, and motional freedom of the object. NMR also provides a tool for examining hydrogen bond geometry. Recent measurements have confirmed the straighter/stronger hydrogen bonds characteristic of cold (more highly structured) water.(25)

In the 1970s, Dr. Raymond Damadian began to study the molecular structure of water within the cell. Using MRI (a form of NMR), he determined that cancerous cells contain what he called "disorganized" water, characterized by "increased motional freedom" of the water molecules. According to his research, the difference between healthy and cancerous cells was the organization of the water within the cell, especially significant surrounding the DNA.(8) Dr. Damadian was the first to suggest that MRI could be used to diagnose cancer. His work provided the basis for this widely-used diagnostic tool which routinely measures the organization of water in the cell as a diagnosis for many forms of cancer.

NMR measurements for water are conducted using ^{17}O-NMR where values are plotted as a function of line-width. Narrower line-widths indicate a greater degree of hydrogen bonding, a more stable water environment, and/or a smaller molecular unit. According to measurements made by Dr. Jhon, hexagonally-structured water has a half line-width between 60 and 70 Hz.(17) Bulk water can measure between 85 and 150 Hz., depending on the quality of the water. Most tap water sources today measure 100 Hz. or greater.(17)(31)

Stabilizing hexagonal water

When water is placed in an electric or magnetic field, the molecules align and assume a more organized, hexagonal geometry. When the field is removed, water generally begins to lose its structure and to resume its less organized form. What then, creates hexagonal water with any degree of stability? Can other energetic influences, or a combination of them, establish the degree of coherence required to stabilize water's structure? Some researchers say, "Yes."

Many believe that the influence of vortex energy, which is difficult to quantify, helps to stabilize structure. This idea is connected to the work of Austrian scientist Viktor Schauberger who accomplished fantastic hydrological feats using the principle of the vortex to optimize water's energetic potential.

Consider also that when water molecules are linked together in a network, hydrogen bonds cooperate to keep the network together. In some instances cooperative bonds have been estimated to be up to 250% stronger than the bond holding 2 water molecules together.(20) This, together with any degree of coherence would contribute to water's ability to hold the hexagonal structure.

Millions of years ago Earth's magnetic field was much stronger than it is today. Many believe that the water on the planet may also have been more hexagonally structured at that time. In fact, some believe that Earth's weakened magnetic field and the reduced amount of hexagonally structured water on the earth are related. This would explain why biological organisms respond favorably to the hexagonal form of water. It would also explain why certain symptoms that have been referred to as *magnetic field deficiency syndrome* disappear when individuals are repeatedly exposed to stronger magnetic fields.

The obvious next question is: If Earth's magnetic field were stronger, would the water on the planet naturally maintain greater structural stability? We may never know. Today, water on the planet is exposed to a weak (.5 gauss) magnetic field. The water is also often in such a polluted state that it is unable to assume hexagonal geometry even when it freezes. This has been illustrated by Dr. Emoto in his "*Messages from Water*" books.(10) Both pollution, and a weakened planetary magnetic field may account for a lesser degree of hexagonal structuring in our water today.

HEXAGONAL WATER AND BIOLOGICAL FUNCTION

Water is a nutrient – ultimately more important than other nutrients since without water even the best vitamins and minerals cannot be absorbed or made available for cellular processes. The *Healing Food Pyramid* recently released by the University of Michigan Integrative Medicine Clinical Services(40) depicts water at the bottom of the food pyramid acknowledging it as the foundational "food" for our health.

Life has evolved around water. The very fact that we are mostly water is evidence of the sophisticated relationship between water and biological systems. Water is involved in every biological function and also plays a significant role in aging.

Aging

In 1986, at a symposium on cancer, Dr. Mu Shik Jhon proposed an entirely new theory for aging. His theory, which he called the *Molecular Water Environment Theory,* states the following:

> *Aging is a loss of hexagonal water from organs, tissues and cells, and an overall decrease in total body water.*

Since the proposal of the above theory, the role of water in aging has been the subject of much research. One scientist, Dr. Seiji Katayama, used magnetic resonance imaging (MRI), to test the hypothesis that aging is related to structural changes in intracellular water. In his study, the structure of the cell water components of four generations within the same genetic family was analyzed. He concluded that age *was*, in fact, related to structural changes in the water surrounding biological macromolecules, and he suggested that "the state of the water in the body [was] closely related to the aging mechanism."(18)

Today there is considerable evidence that rejuvenation may be as simple as re-hydration. According to Dr. Mu Shik Jhon's theory, "Replenishing the hexagonal water in our bodies can increase vitality, slow the aging process, and prevent disease."(17) On a cellular level, there are two water-related factors that are closely associated with aging.

1. The amount of water in the human body declines with age. Newborn infants are over 90% water – by weight. However, by the time a baby is a year old, s/he is only 70% water; it is not uncommon for the elderly to be less than 50% water.

2. The movement of water in and out of cells slows with age. As water movement (turnover) slows, the ratio of water inside the cells to outside the cells is dramatically affected. Cells function best when 60% of the total body's water is inside the cells. However, with age, the amount of water inside the cell gradually decreases – often measuring as low as 40%. This reverses the osmotic potential at the cell membrane and makes it very difficult to re-hydrate the cell. Perhaps this is one reason why it takes longer and is more difficult to re-hydrate an elderly person.

Live blood analysis

Live blood analysis using phase-contrast microscopy offers the unique ability to view the blood in real time. It is being used by a large number of health practitioners worldwide. By providing a dark background, red blood cells and other components of the blood become clearly visible – much like stars against the night sky. Trained technicians and practitioners are able to note the presence of parasites, fungi, bacteria and other conditions of the blood that help to determine a course of action for healing as well as for monitoring the process.

One of the most common conditions found during live blood analysis is a condition known as "Rouleau," where red blood cells stick together. Individuals with Rouleau often feel tired and tend to over-consume proteins and fats; this leads to increased fatigue and weight gain. The stickiness (agglutination) of red blood cells lowers oxygen exchange in the tissues as well as nutrient utilization, cellular efficiency, and the removal of wastes. Agglutination in the blood creates sluggishness in both the circulatory and lymph systems with an obvious decrease in immune function and peripheral blood circulation. Individuals with ongoing Rouleau are ideal candidates for illness and disease.

According to Biokineticist Stefan Schoeman, drinking normal water has little effect on Rouleau in the short term. However, drinking just 16 ounces of hexagonal water produces visible improvement within minutes – sometimes as much as an 80% improvement. He reports the following:

> *In my experience, it takes a serious effort over a period of time to re-hydrate the body and the blood. With that in mind, I have been extremely surprised to discover that within a very few minutes after drinking hexagonal water,*

the Rouleau effect begins to resolve. I believe in a total change in lifestyle; becoming alkaline requires changes in diet and drinking hexagonal water daily.(32)

The photos in Fig. 3 illustrate rapid resolution of Rouleau after drinking 16 ounces of hexagonal water. In the photo on the left (before drinking hexagonal water), red blood cells are stacked tightly together almost like rolls of coins. Rouleau limits the surface area of red blood cells which also limits nutrient and oxygen exchange. The photo on the right (after drinking hexagonal water) shows a separation of the red blood cells and the capacity for greater oxygen uptake and more efficient nutrient utilization.

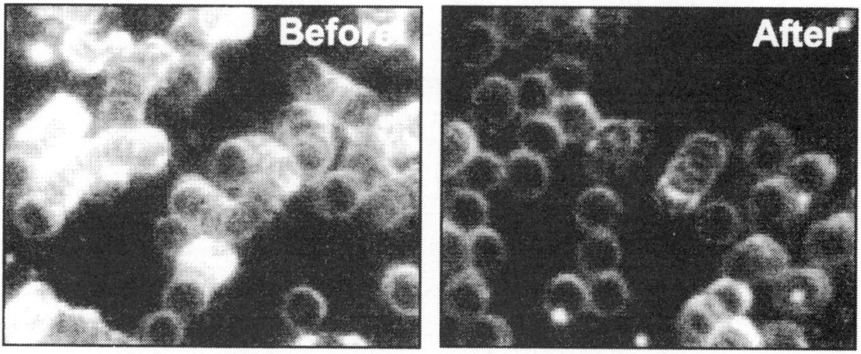

Fig. 3 Live blood samples – before and 10 minutes after drinking hexagonal water

Photos: courtesy of Stefan Schoeman, Biohouse clinic, South Africa

Heart Rate Variability (HRV)

Heart Rate Variability (HRV) testing is a non-invasive, medically approved, computerized system designed to access the state of health in an individual by determining the stress on the autonomic nervous system. Readings are so accurate they can predict heart attacks to within an hour. In preliminary testing, HRV scores improved (up to 30%) within a short period of time following the consumption of hexagonal water. Dr. Bruce Shelton, MD MD(H) DiHom FBIH had this to say of the results of his testing:

> *Testing with devices designed to accurately measure the state of health in an individual, clearly show that water treated with [this device] improves a number of health parameters – within minutes. Tests conducted in our office, before and after drinking 1, 8-ounce glass of water, showed significant improvement. Heart Rate Variability (HRV) scores improved showing an average of 29.4% improvement in a 5-10-minute period of time. As indicated by the above, this is clearly a positive step towards achieving better health. It is a significant addition to a complete program of health care and maintenance.*[35]

Protein

Organized water is recognized as essential for many biological functions.[33][44] It has even been suggested that pressure waves formed by flickering water clusters may link protein molecular vibrations carrying information through the intracellular components of the body.[42]

Proteins have no activity outside the presence of water. They must fold in specific three-dimensional configurations in order to be biologically active. This is accomplished around water which contributes both energy and stabilization. The first layer of water surrounding healthy protein is highly structured water.[17][47] Healthy proteins are surrounded by the highest number of hexagonally-structured water molecules; cancerous cells have a decreased number of hexagonal structures and an increased number of pentagonal structures.[17] Recent research also shows that the first layer of water surrounding healthy proteins has a high proton transfer rate.[47] This is another indicator of structure.

DNA

Water is integral to the structure of DNA. In fact, early computer simulations of DNA completely failed until water was included as a functional part of the simulation.[14] It was later discovered that water interacts with every part of the DNA's double helix. Structured water is required for the precise replication of the helical strands.[16] It stabilizes the entire helical configuration by forming hydrogen bonds with the phosphate groups.

As with other proteins, the first layer of water surrounding healthy DNA is organized water. This stabilizes and protects DNA from energetic influences that could cause malfunction and distortion. Using computer simulation, Dr. Jhon determined that one of the differences between normal and abnormal DNA was the number of water molecules in the first hydration layer. He found a greater number of water molecules around normal DNA than around abnormal DNA. He also found that water molecules were less mobile (highly structured) around normal DNA but were organized to a lesser degree around abnormal DNA.[17]

Intracellular water

The water inside the cell is structurally different from the water outside the cell. As early as 1989, it was proposed that there was a two-phase structure of water where the coherent phase (characterized by molecules that rotated together), accounted for the electrodynamic properties of water, while a fluctuating phase (characterized by molecules that moved randomly) accounted for the thermodynamic properties of water.[9]

Phillipa Wiggins and others have suggested that there may be alternating micro-domains within each cell where both low density (hexagonal) and high density (bulk) water play a critical role in maintaining the balance of solutes within the cell.[(43)] This theory is based on the fact that certain ions (called structure-making ions) tend to support the structuring of water and that certain other ions (referred to as structure-breaking ions) actually discourage the structuring of water.

Potassium (K^+) is a structure-breaking ion which is known to be highly concentrated within the cell. It naturally gravitates toward hexagonal water and then causes the dissolution of the hexagonal structure. On the other hand, Sodium (Na^+) is a structure-making ion. It is found in large concentrations outside the cell. Wiggins' work showed that structure-breaking ions create local osmotic pressure gradients which are abolished by the conversion of hexagonal water to its less structured form. This could be the mechanism (contrary to accepted theory) responsible for the sodium/potassium ion exchange at the cell membrane.[(45)]

A source of energy

Water is the body's main source of energy. According to Dr. Batmanghelidj, author of the well-known book, *Your Body's Many Cries for Water*, the flow of water through the cell membrane provides electrical energy much like the turbines in a hydroelectric plant.[(2)] Hexagonal water has the capacity to hold more energy than normal water because of its structure. It also has the capacity to transfer more energy (see Thermal conductivity & Specific heat, pgs. 11-12).

Only when there is sufficient water do we have ample and ongoing energy, especially as we age. The consumption of hexagonal water often enhances the amount of energy available in the body as explained by the above and as evidenced by the increased energy felt by many who consume it. Drinking a glass of hexagonal water in the afternoon entirely eliminates afternoon fatigue for some individuals.

Hexagonal water may also have effects on energy storage and endurance. A recent study showed that mice who consumed a hexagonal water product were able to swim 47% longer prior to exhaustion and that they had greater stores of glycogen (a source of stored energy) in the liver.(5) This has tremendous implications for the endurance athlete.

Nutrient absorption

The digestive process requires copious amounts of water in order for nutrients to be adequately assimilated. Anything that enhances water movement in the body also enhances nutrient absorption. Hexagonal water is considered an excellent carrier for nutritional supplements. It is being used by a number of athletes and athletic teams to improve nutrient absorption, hydration, and endurance.

Immune function

Hexagonal water also appears to enhance immune function. In a recent study, mice were given a hexagonal water product while a control group was given de-ionized water. After just two weeks, cell culture analysis of the spleen (an organ directly connected with immune function) indicated immune modulation.(30) Perhaps this is one reason why individuals who drink hexagonal water repeatedly report better overall health.

HEXAGONAL WATER FOR VARIOUS AILMENTS

Imagine the static on a radio when there is interference in the atmosphere and the radio message becomes distorted. It is the same within our bodies when the water loses coherence and becomes less organized. Since water carries the vibrations and frequencies that transfer information, signals and messages can be distorted when the coherent network is disrupted. When hexagonal water is repeatedly consumed, greater balance is often re-established as coherence in the water network is enhanced. Hexagonal water is being heralded for its beneficial effects on a variety of health conditions, particularly those that are associated with dehydration.

Dehydration

Most diseases can be traced to dehydration. When there is insufficient water, every bodily function suffers and the body must operate from a level of "crisis management." Ultimately, crisis management results in disease. This has been well documented from the medical perspective, by Dr. Batmanghelidj.[1][2] It is also well known in the field of energy medicine; practitioners agree that it is difficult to work on a dehydrated patient. Therapists using energy/vibrational medicine experience much faster results when their patients are fully hydrated.

Many symptoms experienced by large numbers of the population are ignored for what they are – signs of dehydration. These symptoms are experienced by so many people that they are often considered "normal." Such symptoms include acne, afternoon fatigue, constipation, dry skin, heartburn, and headaches.

When hexagonal water is consumed, a rapid resolution of the above and of other symptoms of dehydration is often noticed.

Since water movement into the cells is accomplished one molecule at a time,[18] structural organization becomes a key factor. For example, the carpenter who uses a nail gun where nails are delivered in an organized string can accomplish his task much quicker than the carpenter who has to reach into a box for each nail. Drinking organized, hexagonal water provides superior hydration as shown in a recent preliminary study.[12] This can have a positive effect on numerous health conditions.

Dry skin, acne and premature wrinkles

The skin is the largest detoxification organ in the human body. It requires a continual supply of water as evaporation and perspiration continually pull water through the skin to eliminate wastes and cleanse the tissues. Water keeps the skin plump and smooth. Without adequate hydration, wastes build up, causing acne and other skin problems.

Because the skin is a reflection of the inner body, it is often the most obvious indicator of dehydration. Dry, flaky skin is a sign that the water level in the body is inadequate. Further symptoms, such as acne and premature wrinkles, are sure signals that the body lacks sufficient water.

One of the most frequently noticed benefits of drinking hexagonal water is a rapid improvement in the condition of the skin. Many notice softer skin within days. Because of its ability to move with greater ease into the cellular components of the body, hexagonal water can often do more for the skin than lotions and creams which clog pores and hinder the cleansing process. Susan experienced marked differences in the condition of her skin in just a few weeks:

> *...This "living water" has also made remarkable improvements to my skin – not just on the face, but on the*

body too. I have had problems with my skin all my life, but in just six weeks my adult acne has cleared up – and is completely gone. The hydration is amazing which has resulted in fewer wrinkles and a very healthy glow. My hair and nails are growing faster and appear to be much stronger.

When I first began to drink hexagonal water, I noticed a definite detoxification – evidenced by a change in bowel movements. Then I noticed that my skin seemed to have "plumped up" which resulted in many of the small lines from my face (especially around my eyes and forehead) literally disappearing...

As a person involved in the health and wellness industry for decades and also one who works with harmonic vibration, this is clearly one of the most remarkable products I've ever seen. This six-sided, hexagonal, living water, in my opinion, is going to have a massive impact on the health and well-being of the whole world. This product is in a class all its own!

Susan S. – Canada

Constipation

The widely accepted statement, "health begins in the colon," might be more accurately modified to read: "health begins with a *hydrated* colon." Water is the key. When the body is forced to remove solid wastes without an adequate supply of water, constipation is the result. Because the major elimination systems of the body (skin, kidneys and bowels), are critically dependent on water, they are primary indicators of drought. Like dry skin, constipation and dark urine indicate water shortage. Hard, infrequent, and difficult-to-eliminate bowel movements indicate that the body is rationing water.

Dr. Mu Shik Jhon and Dr. Gyu Hwan Choi of the National University School of Medicine in Korea conducted a clinical study to determine the effects of hexagonal water on constipation. Their results showed that within four weeks, 75% of patients with chronic constipation had an increase in bowel frequency. Their study documented a 40-60% improvement in transit time through the colon, while control subjects (those with regular bowel movements) who also drank hexagonal water showed no change.(17) This study provides scientific evidence substantiating the often-reported improvement in constipation for those who drink hexagonal water. Dan's experience below is not uncommon:

> *...Several days after I started drinking hexagonal water I experienced definite changes in my elimination patterns. Now after three months, I have come to accept the fact that three bowel movements a day is normal.*
>
> *I have experienced several other pleasant results from drinking this structured water. First of all, I used to have terribly dry skin – especially on my forehead and scalp. That has disappeared. No more dandruff and no more flaky patches on my forehead...*
>
> *I recently left the country and was without hexagonal water for two weeks. For the first week everything was fine – but by the end of my trip, the dryness on my forehead and scalp was returning. I was certainly glad to get back to hexagonal water again – and the dryness went right away.*
> *Rev. Dan J. – Canada*

High cholesterol

Cholesterol is a naturally-produced, waxy substance that is

found throughout the body. It is a functional part of cell walls and a building block in the production of various hormones and Vitamin D. While the build-up of cholesterol in arteries can be harmful, mounting evidence suggests that cholesterol is the body's solution to a bigger problem – dehydration.

It was Dr. Batmanghelidj who exposed the relationship of dehydration to cholesterol.(1) His attention to the water mechanisms of the body has shown that it is dehydration that causes the build-up of cholesterol in arteries and cell tissue – not the cholesterol from foods.

When water is being rationed, cells deposit cholesterol in the lipid bi-layer of their membranes to protect the cell from water loss. This appears to be a natural consequence of aging – perhaps because the thirst mechanism declines with age and elderly people tend to drink less water.

But why does cholesterol build up in our arteries? According to Dr. Batmanghelidj, this is a consequence of concentrated (thick) blood which results from the digestion of food when too little water is available. Concentrated, acidic blood not only triggers the production of cholesterol, it also causes microscopic abrasions in the lining of arteries. Cholesterol covers these abrasions in an effort to repair the damage.(1) Unless water intake is increased, elevated levels of cholesterol continue to circulate in the blood, adding to the build-up in arteries and cell membranes – all because of dehydration.

The good news is that cholesterol build-up can be reversed. With an increase in water and improved water turnover, the blood is naturally thinned, high blood cholesterol is no longer necessary, and reduced cholesterol levels are found in the blood. In time, cholesterol build-up in the arteries can literally be washed away. The reduction of blood cholesterol levels is illustrated in the following account:

Several years ago, I began to have chest pains. Though my situation was not life-threatening, I was told that my cholesterol levels were very high. Ever since then I have struggled to get my cholesterol down. It has been as high as 245, and the lowest it had ever been was 221.

Then a friend of mine introduced me to hexagonal water and I began to drink it. The results were amazing! After only six weeks, I returned to my doctor for more blood tests and was told that my cholesterol was 182 – the lowest in seven years! What five years, three types of medication, and hundreds of dollars in expenses could not accomplish, drinking hexagonal water had done. Now, after a year on the water, my cholesterol is still low.

I appreciate the extra boost of energy that hexagonal water gives me and I love the taste. Everyone in the family notices the difference when we have to drink bottled water.

Luis A. – Texas

High blood pressure (hypertension)

High blood pressure has also been correlated with a reduced amount of water in the blood, and with aging. Adults have a lower concentration of water in their blood than children; the elderly have even lower concentrations. This is one reason (accurate or not) that higher blood pressure has been considered "normal" as we age.

The water content of the blood is critical. We have noted that thickened blood results in higher cholesterol levels; it also results in high blood pressure. This is because blood is mostly water. Normal levels of water in the blood are considered to be 80-85%, although some feel that these accepted levels are much too low. Dr. Batmanghelidj suggests that fully hydrated blood should be 94% water.(2)

When there is a reduction in the amount of water in the blood, the total blood volume is reduced and more pressure is required to pump it into the extremities of the body. As the body becomes increasingly dehydrated, more and more pressure is required and blood pressure continues to rise.

Obviously, there are other factors that contribute to high blood pressure; stress is at the top of the list. However, dehydration is the number one cause of stress in the human body. It is well known that stress, regardless of the cause, produces the same physiological symptoms. Water is so important that it alone can often bring blood pressure within normal levels. Water is also the best thing to take during any kind of stress.

Ken's story is an example of how hexagonal water appears to have brought his blood pressure into the normal range and supported his health following open heart surgery:

> *Three years ago (Feb. 2002), I had heart surgery – a quadruple bypass. For a little over a year I followed the standard medical protocol, taking medication for blood pressure and cholesterol which were both still high. About a year after the surgery, I was introduced to hexagonal water. As I began to drink hexagonal water, I noticed an almost immediate improvement in my energy – and I began to lose weight. The longer I drank the water, the better I felt, and I wondered if the water was helping my heart.*
>
> *When my medication was nearly gone, I was told that I needed to take a stress test before my doctor could prescribe any further medication. I was feeling so good that*

I decided to find out what would happen if I stopped taking medication and continued to drink hexagonal water.

A year later, I went in for my annual physical. To my surprise, my blood work was perfect. Every one of the 40 parameters that was measured was within the normal range. I was really excited and within several months I scheduled the stress test with my cardiologist.

I wish I had a video of everyone's faces as I effortlessly completed that test! At each higher level of the test, the nurses and technicians asked me if I was okay to go on. For the final three minutes, as I was almost running, they were all standing there in disbelief. My blood pressure stayed well within the normal range (130/80 to start and 160/80 at the end) and I was not really winded when I was done.

When my cardiologist came in to review the results, he had to look twice to make sure of what he was reading. Then he asked the technician, "How long has he <u>not</u> been taking any medication?"

Needless to say, I got a clean bill of health where my heart was concerned and although the doctor was obliged to tell me that it would be in my best interest to resume taking medication for blood pressure and cholesterol, I have decided not to take those medications. I can't believe how good I feel.
Ken M. – Texas

Weight Management

Ken's story illustrates another often-noticed benefit of drinking hexagonal water – weight loss.

Being fully hydrated is a prerequisite for weight loss. When the body is in a state of crisis, (dehydrated) it cannot let go of fat. The only way to convince the body to metabolize fat is to eliminate the crisis so that the body feels *safe* in letting go of unnecessary calories. Hexagonal water may be the most critical component of any *natural* and *permanent* weight loss program.

There are several facets to the hexagonal water-weight interaction:

1. Cells that are continually depleted of the hydroelectric energy supplied by water have to depend on energy from food. This is why hunger and thirst are often confused.[(2)] It is also the reason that overweight individuals are usually more dehydrated. Hexagonal water can provide the energetic equivalent of food, helping to reduce caloric intake.

2. Lack of water causes an increase in fat deposits for several reasons. When water is limited, excess wastes are unable to be flushed from the body. Although the number of fat cells in the body appears to be constant, fat cells continue to expand; they are used as repositories for the excess wastes and toxins that cannot be eliminated from the body.

3. When the kidneys do not have enough water, they must rely on the liver to help with the removal of wastes. Since the liver is responsible for the metabolism of fats, inadequate water forces fats to be stored while the liver helps with the removal of wastes.

According to Christy, hexagonal water not only helped her to lose weight, but it also seemed to awaken a consciousness about her body. She naturally changed other things in her lifestyle that contributed to better health:

... After about a month on the water, we [my husband and I] were both noticing results. First, we slept more soundly and we both felt more refreshed in the morning. Second, we found we were drinking more water – without trying. (For someone who used to have to force four glasses of water down every day – that's saying something!) It was as if our bodies made us more alert to the need. Third, we found that our bowel movements were regular again. Lastly, we can honestly say that we feel younger. We were truly amazed, because we are not unhealthy people. Then my daughter pointed out that the skin on my hands no longer stayed up when she pinched it, but flattened out quickly.

What surprised me the most was an inner desire to start doing "right" by my body. I changed my eating habits and started on raw foods, nuts and grains...and I LOVE IT! After six months, I have lost 30 pounds!
Christy G. – New York

Edema (water retention)

Sometimes individuals who are trying to lose weight experience water retention (edema). Water retention is responsible for additional weight and a puffy appearance. This is most often the result of toxins in the body and not enough water to remove them. Diuretics cannot solve the problem because it is caused by dehydration.

Edema is an excessive amount of water held in the interstitial (outside the cells) tissues of the body. It is a crisis management state where the body holds onto salt in order to maintain water – even though that water cannot move into the cells where it is needed. (The sodium in salt controls the amount of water outside/between cells.) In cases of dehydration, people often have cravings for salty food to help retain as much water as possible.

One of the things that often happens when hexagonal water is consumed is a loss of retained water – and weight. The first thing many people notice is more frequent urination. As hexagonal water hydrates the cells, less efficient water is moved out of the body, carrying excess salt and toxins with it. Swelling and puffiness then disappear, as illustrated by the following example:

> *When I first began to drink hexagonal water, I was having trouble with a lot of swelling in my legs. Within two weeks, it was gone and I noticed that my circulation seemed to be much better. I also began to notice how much more energy I had. I actually felt like doing my work!*
>
> *This water has also helped with my cholesterol. I remember going to the doctor about two months after I began to drink the water. My cholesterol had gone down from 250 to 195 and it has stayed down for the last nine months – ever since I began to drink hexagonal water. Even my doctor told me to keep drinking this water!*
> *Jeanette H. – Texas*

Fatigue

Water provides energy for every system in the body. This refers to the water we drink – not the water that is already tied up in other functions. Adequate and regular water intake insures that water is constantly turning the hydroelectric pumps at every cell membrane carrying nutrients and removing wastes on a continual basis.

Hexagonal water provides abundant, clean energy. Since hexagonal water appears to move more efficiently because of its structure, it is able to provide more energy for cellular functions; all it needs to do is "move" through the cells. Hexagonal water provides noticeable energy for many, especially when they first begin to drink the water. According to Kurt, hexagonal water challenges even the best "pick-me-up" drinks on the market:

> *At the time I started drinking hexagonal water, I was spending $300.00 a month on vitamins. For me, hexagonal water has had a much more profound effect on my energy than those $300.00 worth of vitamins ever did! I keep a gallon jug of water at the office and I probably drink a half gallon throughout the day. I've quit drinking diet sodas and I no longer need my 2:00 p.m. Red Bull™ drink.*
> *Kurt C. – Texas*

Diabetes

In Traditional Chinese Medicine, diabetes is referred to as the "thirst disease." It is typified by increased thirst – a huge signal from the body.

The brain is 85% water – its primary source of energy. But when water is limited, the brain calls for the next available energy source – sugar. And since insulin is not required to carry sugar across brain cell membranes, the call for sugar is an easy way for the brain to get needed energy. Unfortunately, this raises the level of sugar in the blood throughout the body.

The activity of insulin itself appears to be governed by water. According to Dr. Z.Y. Wang, Director of the Cluster Biology Lab at the New Jersey Institute of Technology, "Proper hydration is critical for people with diabetes because water has to be organized around insulin – for every one insulin molecule, you need 440 water molecules."(7) Recent clinical studies with diabetic patients showed that the consumption of a hexagonal water product produced a trend towards lower blood sugar levels in just four weeks.(41) Vanessa's experience is not uncommon for diabetics who consume hexagonal water:

> *I am diabetic (diagnosed with juvenile diabetes at ten years old). I have been on insulin for 18 years and during that time I have tried a number of things to help reduce the amount I take. Several things have helped slightly, but nothing like hexagonal water! Over the course of one month, as I monitored my own blood sugar levels (four times a day), my need for insulin gradually dropped. I have been able to reduce the amount of insulin I take – by over half! I used to take 40 units per day, now I consistently require 15 units. I also used to feel constantly dehydrated – even though I drank lots of water. I was continually thirsty and nothing would quench my thirst. Hexagonal water quenches my thirst! For the first time in a long time, I am not perpetually thirsty.*
> *Vanessa G. – California*

Asthma

Normal respiration places a continual and ongoing drain on the available water in the body. Air sacs in lung tissue must be continually lined with water or they dry out. During extreme cases of water shortage, the body's airways are constricted to reduce water loss; this is asthma. Asthma and many allergies are indicators of dehydration.

One of the reasons children experience the symptoms of asthma more than adults, is that children are continually growing. Each time a new cell is created, it must be filled with water. This puts a drain on the water reserves in a child's body.[(2)] Children, although they are young and resilient, need water just as much (if not more) than adults and the elderly.

Asthma medications open constricted airways. They also shut off the body's call for water, ignoring the bigger problem. In the long run, water is possibly the best medication for asthma and many allergies,[(1)(2)] as illustrated in the following account.

I just wanted to let you know what a profound effect the hexagonal water has had on my health. At the time I started drinking hexagonal water, I was experiencing several medical problems. My blood sugar was unsettled at about 270, I was taking two, 850mg. tablets of Metformin™ a day. My asthma was also giving me trouble and I needed four medications from an asthma pump – daily. I was also taking blood pressure medication three times a day.

When I began to drink hexagonal water, I noticed a difference the very first day. For a few days, I experienced a cleansing – then I began to notice other things. Besides a greater amount of energy, the swings in my blood sugar began to level out. After four months, my doctor reduced the amount of Metformin™ I was taking – to only 1 (500 mg.) pill a day. My blood sugar was down to 100! I was also able to reduce my asthma medication to one pump a day and my blood pressure medication was reduced from three pills to one pill a day. My doctor told me to keep drinking hexagonal water.

Today, after six months, I returned from the doctor with normal blood sugar and I was told that I no longer need any medication for diabetes.

Since I began drinking hexagonal water, I have found myself doing things that I have not done for a long time. I feel so much better. My skin is not as dry and it is noticeably softer.

Rhonda R. – Texas

Mental clarity and cognitive function

Brain functions require electrical activity which is supplied by water. Perhaps this is why the brain has such a high concentration of water. Even though the brain is a comparatively small part of the body, it is allocated 20% of the total blood supply, supplying water, oxygen, and other nutrients to the brain on a priority basis.

Since water movement creates hydroelectric energy, the supply of free water to the brain is essential on a regular basis. When it is not continually available, thinking becomes clouded and the ability to focus becomes difficult. Anyone who experiences brain fog or the inability to concentrate, can benefit (often immediately) from hexagonal water. Dr. David Stokesbary, an orthopedic surgeon and head of the Advanced Magnetic Research Institute in California, comments on this and other benefits of drinking hexagonal water:

Anything that encourages this kind of detoxification is capable of playing a huge role in generating cellular energy. Cellular water movement is critical for adequate absorption of nutrients; also for the elimination of metabolic waste products. Those who drink this water are benefiting in both ways. With better nourishment and more effective waste removal, every process in the body will happen more efficiently. This efficiency accounts for the greater levels of energy which are often experienced. It also accounts for improvements in mental clarity and the heightened sense of well-being so often experienced. And of course, it supports the healing environment.(34)

Joint Pain

When an old door becomes stiff and creaky, we know it is time to oil the hinges. In order to open the door more freely, we have to remove the debris and lubricate the hinge. The joints in our bodies are no different. Cartilage surrounds and protects the bones in our joints. It is like thick Jello and it is mostly water. When there is enough water in the body, joints are lubricated and glide smoothly. However, when cartilage dries out there is friction and greater abrasion as joints move.

Because of the constant shearing forces, cartilage cells must be replaced more often than other tissues in the body. Without adequate water, cartilage breakdown occurs rapidly and new cell development is restricted. Besides the more rapid breakdown of cartilage, too little water limits the body's ability to flush acidic wastes which settle in the joints.

How do we remove the debris and lubricate the joints in our bodies? Once again, the answer is water. Hexagonal water often provides a reduction in joint pain and increased mobility for those with arthritis and other joint problems. In a short period of time, Betty noticed considerable improvement in her ability to stand and walk:

Before I discovered hexagonal water, the arthritis in my knees had gotten so bad that I rarely did any walking. Getting up from a chair was difficult unless I had assis-

tance. I could not stand and talk to people – I had to sit down. Within several weeks after I began to drink hexagonal water, the pain in my knees was noticeably reduced. Eight months later I was able to walk up and down the stadium stairs at a graduation ceremony. Within a year, I went to a state fair and walked through all eight buildings – with no pain at all.

I have been drinking this wonderful water for four years and at the age of 81, I still experience very little (if any) pain in my knees.

Betty E. – California

Emotional Release

Beyond the purely physical, there are other possible benefits from drinking hexagonal water. Howard Wolin, M.D., a practicing physician specializing in advanced nutrition, alludes to the emotional release some have experienced when drinking hexagonal water. For example, he states, "Clinically, I have been able to determine that this water is capable of reawakening dormant and compromised functions in the cells of the body and rapidly eliminating deeply held toxins... even deeply held emotional traumas that have remained hidden. It has the ability to help the body in ways that science has not even begun to understand."[(46)]

Hexagonal water may actually be cleansing on more than one level. It may hold the secret to releasing emotional toxins (in the form of traumas and negative emotional patterns). Many believe that the consumption of hexagonal water can be instrumental in raising the vibratory frequency of those who consume it regularly, in other words, raising consciousness.

Brianna is not alone in believing that her introduction to hexagonal water began a process of cleansing – both physical and otherwise:

Beyond the physical benefits, I have experienced other manifestations of the efficacy of hexagonal water. Within days I began to have intense and very spiritual dreams. On more than one occasion I was surrounded by water – flowing with it and moving through it effortlessly. It was almost as if I was being introduced to water on more than one level. For me, it was additional verification that I was pursuing a correct course of action with the water.

Amazingly enough, I also noticed that shortly after I began drinking hexagonal water, a number of "not so useful" patterns from my childhood, began to surface – allowing me to "look at" and "release" old beliefs and familial patterns that were not in my best interest. I feel like I am having a "love-affair" with this water!
Brianna C. – California

Sources of Hexagonal Water

Natural sources

There are a number of natural springs on the planet where the water is said to have hexagonal molecular structure. Many of these places are known as healing springs; some have been considered sacred by native cultures for centuries. Many of these sources of hexagonal water are located in places where unique magnetic and/or geologic conditions exist. Most are located in areas where the water comes from deep within the earth. Some glacial water supplies also claim hexagonal structure. Although there may be numerous sources of natural hexagonal water, the stability of the structure may vary from location to location. In most instances, the water's structure is thought to deteriorate the moment it reaches the surface or leaves the area.

There is one natural source of hexagonal water that appears to maintain structural stability for an exceptionally long period of time. This water, from Japan's Magnetic Mountain, originates in a place where the magnetic field is unusually high. In fact, the entire mountain has been permanently magnetized. The mountain itself is the result of an underwater volcanic eruption which occurred millions of years ago at about the same time that a meteorite shower deposited magnetite within the mountain. The water from this mountain has its own magnetic field when it comes to the surface, measured at nearly 30 gauss.[(29)] The magnetic field of this water declines with time, but never fully dissipates – one reason the water may be stable. This water has been the subject of several clinical studies, supporting its health-enhancing effects.[(13)(27)(28)(32)]

Movement and natural forces

Most people know enough not to drink stagnant water. When left long enough, stagnant water breeds bacteria and disease. But when water is allowed to move freely, it naturally heals itself. In nature, turbulence is the key to healthy water; the only time water is really sensitive to external forces is when it is moving.

Free water (not water forced through straight pipes under pressure) forms layers as it moves – sheets that fold and slide over each other as the water expands and contracts. With each expansion and contraction, water absorbs electrons and energy. Turbulence and vibration tend to open the molecular structure, increasing hexagonal organization and the energy potential of the water.

Nature repeatedly uses the vortex to gather and sustain energy. Everywhere it is evident – from the smallest strand of DNA to the largest galaxy. Simple observation of water's movement reveals the repeated spiral vortexes, the eddies and the whirlpools it creates as it moves. This vortex action cleanses water, releasing vibrational pollution while at the same time ionizing the minerals and adding structure. As layers of water slip past each other, oxygen is taken in, adding to the total purification of water through movement.

Nature uses other natural forces to energize, cleanse, and structure water. These include the earth's magnetic field, infrared radiation from the sun, a variety of minerals that support the structural stability of water, and other natural frequencies and forces that are difficult to quantify.

Making hexagonal water

In order to change the structure of water, it must be subjected to energetic forces and frequencies that cause water to realign in the hexagonal, crystalline lattice. Depending on the method used, the water produced can be many different degrees of hexagonal. Thus the stability and level of coherence can vary widely between products.

A number of "hexagonal" water products are available today – each with unique qualities and characteristics – some better than others. Several products are sold "ready-to-drink," while others are available as concentrates. Also available are vortexing machines, in-line structuring devices, and ionizers that produce hexagonal water in the home. The difference between these products and the hexagonal water they produce can be quite substantial. Regardless of which method is used, once it is created, the hexagonal network is most stable when kept cool and away from electromagnetic frequencies and direct sunlight.

Concentrates

Many have wondered how "concentrated" hexagonal water can produce the hexagonal structure in a greater volume. Hexagonal water is a crystal – in liquid form. Crystals form very rapidly in saturated solutions – especially when "seed" crystals are implanted. Hexagonal water "concentrates" act as a "seed" or pattern, helping to spread the same structure and frequencies throughout the whole. Since hexagonal water has an enhanced ability to hold and transfer energy, it can readily transfer those same structural qualities and imprinted frequencies to a greater volume.

Hexagonal water concentrates vary widely. Some are imprinted with frequencies to support mental clarity, creative thinking, immune function, cellular regeneration, etc. Those who sell concentrated hexagonal water recommend dilution factors to maximize the efficacy of their particular products.

Vortexing machines

Spiral movement is a prerequisite in all living systems. It is present in the natural flow of water in rivers and streams, in the human circulatory system, and in the planetary movements in space. In nature, countless vortexes form as water breaks past rocks and other obstructions. These vortexes act as powerful resonant structures, energizing, purifying, and ionizing the water as it moves.

The inside layers of water in a vortex flow much faster than outer layers. In a vortex, the speed multiplied by the radius is a constant. Theoretically, this means that the velocity at the center of a vortex is infinite. As the forces in a vortex approach infinity, and as layers expand and contract, hydrogen bonds in the water molecule begin to stretch. This opens the molecular structure of water for a constant interchange and exchange of electrons. As hexagonal structuring increases, the water's ability to carry vibrations and frequency information also increases.

Vortexing machines use a combination of counter-rotational magnetic fields, structure-making ions, and nature's spiral vortex to create hexagonal water. These devices are small enough (about the size of a blender) to sit on a kitchen countertop. They can provide an abundant supply of hexagonal water for family, pets, and plants.

In-line devices

Several in-line hexagonal water-making devices are now on the market. These are designed to be plumbed into the main water line or used as attachments at a water exit (e.g. garden hose or faucet). Most of these devices are designed with fins or spirals inside the water pipe creating vortexes as water flows past. Some devices use counter-rotational vortexes and a variety of minerals inside the casing to add the stabilizing/energizing influence of mineral frequencies to the water.

Ionizers

Ionizers use the principal of electrolysis to separate the ions in water. Positively-charged ions gravitate toward a negative electrode, while negatively-charged ions move in the opposite direction. This creates two forms of water – acidic water and alkaline water. The alkaline water is recommended for consumption while the acidic water has many therapeutic uses as a topical solution. Ionized water is hexagonally structured, however the quality and stability of ionized water are entirely dependent on the equipment being used and on substances that are present in the water.

Begin with "clean" water

Regardless of the method, it is advisable that water be clean and free of contaminants prior to structuring. This requires either the use of a water filter or the purchase of pure water. Some of the more pristine mineral or artesian waters are excellent. When unsure, the use of distilled or high quality RO (reverse osmosis) water is recommended.

According to Dr. Wolfgang Ludwig, there are several methods for eliminating vibrational pollutants. The first and easiest method is turbulence – the repeated stirring (vortexing) of the water. The second method is to treat the water with magnetic energy and/or lasers. The last is to use x-rays, (this method is not recommend because of the possibility of other complications).[21]

Distillation is also used for "erasing" water's memory (clearing the water of vibrational pollutants). Although the process results in very low energy water, it appears to work well as long as temperatures are high enough. However, when distilled water is activated by hexagonal concentrates or other devices, the frequency is raised, producing a higher energy state and water that is enhanced by crystalline structure.

Drinking Hexagonal Water – What to expect

When and how much

One of the best times to drink hexagonal water is first thing in the morning. After six to eight hours without water, the body's reserves need to be replenished. Drinking hexagonal water hydrates the digestive organs and "primes" them before breakfast. It is a good idea to drink a glass (8-16 ounces) 20-30 minutes before *every* meal. This helps to prevent water rationing and all of its complications.

Drinking before meals has other advantages. It helps the body to distinguish between thirst and hunger. This is important for those who are watching their weight since most people tend to eat less when they drink water first. When water is consumed *before* each meal, individuals also tend to chew more thoroughly rather than washing down their food. This results in better digestion and in greater enjoyment of the meal; it also establishes a closer connection with the food we consume. Most health professionals agree that it is best *not* to drink too much water with a meal because it dilutes the stomach acids that are required for digestion.

Water is best consumed at regular intervals throughout the day. As a general rule, drink half your body weight (lbs.) in ounces of water each day. For example, if you weigh 150 pounds, drink 75 ounces (just over a half gallon) of water every day.

Water is especially important before and during exercise. During strenuous activity, blood volume decreases. This is due to an overall reduction in body water (blood is between 85-94% water). Reduced blood volume diminishes the oxygen to muscles and has a subsequent effect on endurance and performance. Many athletes appreciate the improved hydration hexagonal water provides.

Since water is vital for the uptake of nutrients and electrolytes at the cellular level, hexagonal water is perfect when taken with nutritional supplements. It aids in the more efficient assimilation of nutrients. For the same reason, medication should *not* be taken with hexagonal water. It may have a more potent effect and may be too strong. It has been recommended that any medication be taken either an hour before or an hour after drinking hexagonal water. It is also a good idea to work with a health practitioner when consuming hexagonal water, especially if you take prescription medications. Some individuals have been able to reduce or eliminate certain medications.

The manufacturers of hexagonal water products recommend drinking differing amounts – everything from one, to several glasses of their product each day. While this is often enough to reap many benefits, some individuals find that they want to consume hexagonal water exclusively. If you find this is the case for you, remember to build up to this point slowly. Hexagonal water can have powerful detoxification effects, releasing long-held toxins and often producing a cleansing reaction.

Cleansing reactions

When the human body has enough water and energy it is able to nourish and cleanse itself on a regular basis. Water is often the limiting factor. When water is either *insufficient* or *inefficient,* toxins are "tucked away" and stored in waste repositories throughout the body. Unfortunately, when wastes build up, disease develops.

Hexagonal water can provide a significant boost in both energy and cellular efficiency. For those with an overabundance of toxins, the body will respond by clearing out the garbage – often clogging the hallway on the way to the exit. If the toxic load in the body is too great when cleansing begins, a person may feel like s/he has a cold or flu. Symptoms of cleansing include aches and pains, headaches, fatigue, diarrhea, and/or mucus discharges. Each of these is a sign that the body is doing what it is supposed to do. However, it is always more comfortable if a person can cleanse gradually. Beginning slowly is the key.

It is usually best to begin drinking hexagonal water a little at a time. Anyone over the age of 50 or anyone with health problems should consider a slow introduction – no more than 4-6 ounces a day for the first week, while consuming a normal amount of pure water. If you are young and healthy, beginning with 8 ounces a day should not cause cleansing symptoms.

For those who have access to biofeedback equipment, it may be best to "ask the body." Many health practitioners have methods of determining how much hexagonal water the body can accept without experiencing a cleansing reaction.

If you do experience cleansing symptoms (usually within a week or two), stop drinking hexagonal water until the symptoms pass – then begin again more slowly. It is important to continue to drink half your weight (lbs.) in ounces of water every day, even while you are building up to the desired amount of hexagonal water.

Conclusions

Dr. Albert Szent-Györgyi was an inspiration to many who have contributed to our understanding of water. Others, including Viktor Schauberger and Dr. Mu Shik Jhon have helped as we have begun to demystify the miracle of water. Dr. Szent-Györgyi is quoted as having said, "Since the molecular structure of water is the essence of all life, the person who can control that structure in cellular systems, will change the world." We are now witnessing the truth of that statement.

The next few years promise to be monumental as we uncover more pieces of what Dr. Mu Shik Jhon referred to as the *water puzzle*. We may prove what many already believe – that drinking hexagonal water not only improves hydration, augments nutrient absorption, enhances metabolic efficiency, and facilitates the more efficient removal of toxins, but that it is also instrumental in clearing the congested patterns that have slowed our enlightenment. Hexagonal water may hold the key, not just to our health, but to our ability to move to the next level of consciousness.

When we are ready to "become the water" and to allow our connection with every other member of the human family to support conscious, unified action, we will experience quantum leaps in our ability to create solutions which will heal the water, the planet, and all of mankind. Drinking hexagonal water may help us to rise to that occasion. The *Ultimate Solution* may be as simple as our willingness to drink the liquid crystal known as hexagonal water.

References:

1. Batmanghelidj, F. Your Body's Many Cries for Water. Global Health Solutions, Inc. Falls Church, VA 1992.
2. Batmanghelidj, F. Water: for Health, for Healing, for Life - You're not Sick, You're Thirsty. Warner Books, Inc. New York, NY 2003.
3. Belon P. et al. Inhibition of human basophil degranulation by successive histamine dilutions: results of a European multi-centre trial. Inflamm Res. 1999 Apr;48 Suppl 1:S17-8.
4. Chaplin, M.F. A proposal for structuring of water. Biophys. Chem. 83 (1999) 211-221.
5. Chen D. et al. Enhancement of swimming endurance by oral administration of clustered water. J. Internat. Soc. of Sports Nutr. (2005) 2(1):1-30 Poster #42.
6. Cho, H.C. et al. Understanding all of water's anomalies with a non-local potential. J. Chem.Phys. 107(1997) 7979-7988.
7. CSI Dec. 2004. Press release: Better hydration needed for diabetics.
8. Damadian, R. Tumor Detection by Nuclear Magnetic Resonance. Science, 171 (1971) 1151-1153.
9. del Guidice, E. et al. Proceedings of the NATO ASI on hydrogen bonded liquids at Cargése, France preprint MITH 89/10 Apr. 1989 3-15.
10. Emoto, M. The Messages from Water. Tokyo: IHM press, 1999.
11. Evans, W. et al. Thermal conductivity of ordered molecular water. (2006) Unpublished manuscript. A report sponsored by the U.S. Government.
12. Fenestra Research unpublished study. Preliminary clinical,

blind, hydration study for the evaluation of the Vitalizer Plus. Final report submitted to R-Garden Feb 2007.
13. Galitzer, M. and Ohno Y. A Study on the efficacy of a naturally magnetized mineral water on improving kidney and cell electrochemical stability. Conducted at the American Health Institute on Anti-aging Medicine, Santa Monica, CA. Unpublished.
14. Gerstein, M and Levitt M. Molecules of Life. Computer modeling reveals how water affects the structures and dynamics of biological molecules such as proteins, yielding clues to their functions. Scientific American Nov 1998 101-104.
15. Higo, J. et al. Large vortex-like structures of dipole field in computer models of liquid water and dipole-bridge between biomolecules, Proc. Natl. Acad. Sci. USA 98 (2001) 5961-5964.
16. Iqbal, Jawed, Ph.D. Regulatory Role of Cellular Free Water, Sci. Med. Simplified, Apr 1990 Vol. 1, pp 41-54.
17. Jhon, M.S. The Water Puzzle and the Hexagonal Key Uplifting Press, Inc. 2004.
18. Katayama, S. Aging mechanism associated with a function of Biowater. Physiol. Chem. Phys & Med. NMR (1992) 24:43-50.
19. Knepper, M.A. et al. Peter Agre, 2003 Nobel Prize winner in Chemistry. J. Am. Soc, Nephrol. (2004) Apr 15(4):1093-95.
20. Luck, W.A. The importance of cooperativity for the properties of liquid water. J. Mol. Struct. 448 (1998) 131-142.
21. Ludwig, W. An interview published in Environmental Medicine by Michael Trevan and Peter P Talkenberger available online at: www.theaethergroup.com/water_indepth.asp.
22. Ludwig, W. Acta Medica Empirica, 4(1991) 293.
23. Mishima, O. et al. The relationship between liquid, supercooled and glassy water. Nature 396 (1998) 329-335.

24. Mitsui, T. et al. Water diffusion and clustering on Pd(111). Science (2002) Sep 13;297(5588):1850-2.
25. Modig et al. Temperature-dependant hydrogen-bond geometry in liquid water. Phys.Rev. Lett. 90(2003) 075502.
26. Noriyuki, Y, et al. A molecular dynamics study of dielectric constant of water from ambient to sub-and supercritical conditions using a fluctuating charge potential model. Chem. Phys. Lett. (2001) 345(1-2) 195-200.
27. Ohno, Y. et al. Naturally magnetized water and the stability of bacteria in intestinal flora: Implications for health and anti-aging. Presented at the American Academy of Anti-aging Medicine Las Vegas, NV Dec. 1999.
28. Ohno, Y. The effects of magnetized mineral water on memory loss delay in Alzheimer's disease. Frontier Perspectives - a publication of Temple University (1997) Volume 6 No. 2.
29. Ohno, Yoshitaka, M.D. PhD. - Personal communication. 5/17/2005.
30. Pan, J. et al. Low Resonant Frequency Storage and Transfer in Structured Water Cluster IEEE 2003 SMC proceeding (2003) 5034-5039.
31. Pan, J. et al. Clustered Water and Bio-signal Networks. IEEE 2004 Proceedings.
32. Reminick, H. A study on the effect of naturally magnetized water on increasing cellular hydration and its relationship to health and longevity. Explore 10:6 (2001) 58-61.
33. Robinson, G.W. et al. Role of hydration water in protein unfolding. Biophys. J. 77 (1999) 3311-3318.
34. Schoeman, Stefan - e-mail communication 6/21/2005.
35. Shelton, Bruce H. MD MD(H) DiHom FBIH Statement of findings submitted to Vortex Water Technology Jan 17, 2006.

36. Stanley, H.E. et al. The puzzling statistical physics of liquid water. Physica A 257 (1998) 213-232.
37. Stokesbary, D. M.D. An informal discussion with David Stokesbary, MD The Clustered Solution 2003 vol. 2 pg. 4.
38. Suresh, S.J. & Naik, V.M. Hydrogen bond thermodynamic properties of water from dielectric constant data. J. Chem. Phys. (2000) 113(21) 9727-9732.
39. Tiezzi, E. NMR evidence of a supramolecular structure of water. Ann. Chim. (2003) Apr;93(4) 471-476.
40. University of Michigan Integrative Medicine Clinical Services, Healing Food Pyramid available at: www.sciencedaily.com/releases/2005/03/050309133427.htm.
41. Wang, Z.Y. et al. Microclustered water and hydration. Asia Pac. J. Clin. Nutr. (2004)13: S128.
42. Watterson, J.G. A model linking water and protein structure. Biosystems 22(1988) 51-54.
43. Wiggins, P.M. High and low density water in gels. Prog. Polymer. Sci. 20 (1995) 1121-1168.
44. Wiggins, P. M. High and low density intracellular water. Cell. Mol. Biol. 47(2001) 735-744.
45. Wiggins, P.M. Enzyme reactions and two-state water. J. Biol. Phys. And Chem. 2 (2002) 25-37.
46. Wolin, H. Scientific Health Discovery. The Clustered Solution (2003) vol. 2. pg. 4.
47. Yokomizo, T. et al. Hydrogen-bond patterns in the hydration structure of a protein. Chem. Phys. Lett. 401(2005) 332-336.

Note: *Signed statements from indidviduals quoted in this book are on file with the author.*

About the Author
MJ Pangman, M.S.

MJ Pangman is a writer specializing in complementary medicine. Her education and background in the plant sciences give her an understanding of biology, chemistry, and other scientific fields. At the same time, her ability to explain the complex in simple terms makes her work easy to read. Since the year 2000, ever since she observed its powerful effects on her own body, MJ has been writing about structured water – she acknowledges that it is the only water she drinks. MJ worked with the renowned scientist Dr. Mu Shik Jhon to make his book (outlining over 40 years of scientific research), available in English. That book, *The Water Puzzle and the Hexagonal Key*, is available on Amazon.com.

MJ by the river near her home in Wanship, Utah